*Giggles, Gags and Groaners*

# MATHMAGICAL FUN

WRITTEN BY:

STUART A. KALLEN

Published by Abdo & Daughters, 6535 Cecilia Circle, Edina, Minnesota 55439.

Library bound edition distributed by Rockbottom Books, Pentagon Tower, P.O. Box 36036, Minneapolis, Minnesota 55435.



Printed in the United States.

Cover Illustration: Terry Boles
Inside Illustrations: Terry Boles

**Edited By: Rosemary Wallner**

Kallen, Stuart A. , 1955-
Mathmagical Fun/written by Stuart A. Kallen
p. cm.-- (Giggles, gags and groaners)
Summary: A collection of math tricks using the magic of numbers.
ISBN 1-56239-129-1
1. Mathematical recreations--Juvenile literature.
[1. Mathematical recreations. 2. Magic tricks.] I. Title.
II. Series: Kallen, Stuart A. , 1955- Giggles, gags and groaners.
QA95.K33 1992
793. 7'4--dc20 92-14777
CIP
AC

| **International Standard Book Number:** | **Library of Congress Catalog Card Number:** |
|---|---|
| 1-56239-129-1 | 92-14777 |

# TABLE OF CONTENTS

# EASY AS 1...2...3...

Magicians do amazing things. They pull rabbits out of their hats. They saw people in half. They even make people disappear! Magicians do all sorts of tricks with quick hands, special props, and a lot of practice. Most magicians perform tricks that fool the eye. But some tricks can fool the mind.

When you enter into the world of math-a-magic, you can baffle and mystify your friends. Math magic is done without rubber bands up your sleeves, false-bottom boxes, or phony magic wands. You perform mathematical magic using the magic of numbers.

And if it weren't for numbers you wouldn't have any fun. Almost every game is played with numbers. Card, dice, and board games are almost all numbers. And if it weren't for numbers, you couldn't keep score in games like basketball, football, and baseball.

The reason you can have so much fun with numbers is because numbers always play by the rules. You can add, divide, subtract, and multiply them. You can toss them in odd ways, sprinkle them with word play, and flavor them with jokes. And if you cook them just right, the numbers will always add up.

You can perform the magical math tricks in this book for an audience or just your best friend. Of course, you'll want to practice these tricks before attempting to perform them in front of an audience.

At first, you might need a calculator for some of these tricks. But once you get the formulas down, you'll be able to convince your friends that you're a mind reader. Or better yet a math-a-magician. So fire up your calculator and let the math-a-magic begin!

Note: To demonstrate how these tricks work, we picked numbers at random. But any numbers will work with these formulas.

6
7
8
9
10

# AN EXTRA MAGIC FINGER

This trick is silly but it's sure to get a laugh. Tell all your friends that the reason you're a math wizard is because you have an extra finger.

When they scoff, hold up your left hand. Tell them that you will give each finger a number.

Count down from ten on your fingers. Touch each finger as you go. Touch your little finger and say, "This is finger number ten." Touch your ring finger and say, "This is finger number nine." Touch your middle finger and say, "This is finger number eight." Touch your index finger and say, "This is finger number seven." Finally, touch your thumb and say, "This is finger number six."

Now hold up your right hand and say, "And on this hand I have five fingers. Six fingers plus five fingers equal eleven fingers."

# NUMBERS ON THE TABLE

You need a sheet of paper and a pencil for this trick. On a sheet of paper write the numbers 1 through 9 as in the illustration. Carefully tear the paper into nine pieces with a number on each.

Have a friend place the numbers face down on a table and scramble them up.

Tell your friend that you know all and see all. You can pick the number 5 from the piles.

Find the piece of paper that has four ragged edges. Pick it up and show your friend the number 5 written on the paper. Because 5 is the middle number on the paper, it is the only one that will have ragged edges on all four sides.

Another way to do this trick is to ask your friends whether they want you to pick all the even numbers or all the odd numbers.

If they say "even," pick the pieces of paper with one smooth edge. If they say "odd," pick the pieces of paper with two smooth edges. Of course the 5 will still have four rough edges and that is an odd number. Don't forget to grab the five.

An even trickier way to do this trick is to put all the numbers into a hat. Ask your friends if they want all the even numbers or all the odd numbers.

Then feel around for the pieces with the one smooth edge to pick the even numbers. Feel around for the two smooth edges plus the piece with all the ragged edges for the odd numbers.

## NUMBER ESP

In this trick you can correctly guess two numbers picked by someone else. If the person adds and multiplies the numbers correctly, you can pretend to "read their minds."

Ask someone to write down any two numbers between 1 and 9. Do not look at these numbers.

**Say your friend picks the numbers 8 and 2.**

Tell your friend to multiply one of the numbers by 5.

**8 X 5 = 40**

Tell your friend to add 3 to the answer.

**40 + 3 = 43**

Ask your friend to multiply that answer by 2.

**43 X 2 = 86**

Tell your friend to add the other number that was picked to the sum.

**86 + 2 = 88**

Ask your friend the number. When your friend says

"88," subtract 6. You will get the original two numbers.

**88 - 6 = 82**

The original numbers were 8 and 2!

This trick will work with any two numbers between 1 and 9. When you subtract 6 at the end, you will always get a number that is made up of your friend's original two numbers.

# X - RAY VISION

Using this trick, you can fool your friend into thinking that you can see through their hands.

Tell your friend to pick up one penny and one dime. Tell your friend to put the dime in one hand and the penny in the other. Do not look at which coin is in which hand.

For the sake of explanation, let's say that the penny is in your friend's left hand. That would mean that the dime is in the right hand.

Tell them to multiply the coin in their right hand by 4.

**10¢ X 4 = 40**

Tell them to multiply the coin in their left hand by 5.

**1¢ X 5 = 5**

Tell them to add the two numbers together in their head.

**40 + 5 = 45**

Ask the answer. If the number is an odd number, the penny is in the left hand and the dime is in the right.

Point to the left hand and say, "There's the penny." If the number is an even number, the penny is in the right hand and the dime is in the left. Remember, if you are facing your friend, your right is their left. This sounds a little confusing, but with practice you will amaze your friends.

This trick works because when you add two even numbers you always get an even number. When you add an odd number and an even number you always get an odd number. When you multiply the coin in the right hand, you will always get an even number. When you multiply the coin in the left hand, you will get an odd number if it is a penny or an even number if it is a dime.

## THE 99 TRICK

The answer to this trick is always 99. You might want to give your friend a calculator to make this trick easier for them.

Ask your friend to write down any two numbers between 50 and 100.

**Say your friend picks 72 and 55.**

Tell them to add the two numbers.

**72 + 55 = 127**

Ask them to cross out the digit on the left.

**~~1~~ 27 (that leaves 27)**

Tell them to add 1 to the number.

**27 + 1 = 28**

Ask your friend to subtract that number from the original sum.

**127 - 28 = 99**

The number will always be 99!

## STRING OF NUMBERS

This is a tricky trick that requires a calculator or a piece of paper. Without a calculator, you will need a good knowledge of your multiplication tables.

Pick your favorite number between 1 and 9.

72 +55 =127
+27
27+1=28

4 x 123456789=
49382716
9 x 49382716 =
444444444

**Let's say that your favorite number is 4.**

Now multiply your favorite number by 12345679. Notice that there is no 8 in the large number.

**4 X 12345679 = 49382716**

Now multiply *that* number by 9. (Now you can see why you need a calculator.)

**9 X 49382716 = 444444444**

This trick will always give you a string of numbers that is the same as the number you picked at the beginning. By the way, the number we ended up with is four hundred forty-four million, four hundred forty-four thousand, four hundred and forty-four!

## FUN FOR THE MONEY

Here's a trick where you can guess someone's age and how much money they have in their hand. You might want to give the person a calculator to make it easier for them especially if the person is not a math-a-magician.

First, find someone whose age you do not know. Ask them to remove the loose change from their pocket or purse.

Turn your back to them and tell them to count off any amount of coins under one dollar. Have them put the money somewhere where you cannot see it. They may show the money to anyone else who is watching. Make sure not to look at the money that they are counting out.

Tell the person to multiply his or her age by 2.

**Let's say that the person is 12 years old.**

**12 X 2 = 24**

Tell the person to add 5 to the result.

**24 + 5 = 29**

Tell them to multiply the sum by 50.

**29 X 50 = 1450**

Tell them to subtract the number of days in a year.

**1450 - 365 = 1085**

Tell them to add the amount of change they counted out to the end result.

**1085 + 67 = 1152**

Ask the person to write that number down on a piece of paper and hand it to you. Secretly add 115 to the number on the paper.

**1152 + 115 = 1267**

The first two digits equal the person's age. The second two digits equal the amount of money he or she is holding. You can say, "You are twelve years old and you have 67 cents!" This trick will work with any age and any amount of money under one dollar.

## DOWN THE LINE

You need three people and a calculator for this trick.

Hand the calculator to the first person and turn your back. Tell the person to pick any three-digit number and multiply that by 7.

**Let's say that the person picks 456**

**456 X 7 = 3192**

Tell the first person to hand the calculator to the second person. Tell the second person to multiply that number by 11.

**3192 X 11 = 35112**

Tell the second person to hand the calculator to the third person. Tell the third person to multiply that number by 13 and hand you the calculator.

**35112 X 13 = 456456**

Tell everybody that the number chosen was 456. Whatever three-digit number that is picked will be displayed twice in a six-digit number on the calculator.

## MATH-A-MAGICAL CARD

This is a great way to guess which card your friend has picked out of the deck. Once again, a calculator might come in handy.

CLUBS = 1
DIAMONDS = 2
HEARTS = 3
SPADES = 4

On a piece of paper, draw the symbol for each suit of cards. Write the numbers next to them as in the illustration. Leave this chart where your audience can see it.

Clubs = 1  Diamonds = 2  Hearts = 3  Spades = 4

Hand someone the deck of cards. Tell them to shuffle the cards and pick one. Have them show the card to the audience while your back is turned. Do not look at the card. Tell the person to write the number of the card on a piece of paper. The ace is 1, the Jack is 11, the Queen is 12, and the King is 13. Tell your friend to multiply the number of the card by 2.

**Let's say your friend picked the Jack of diamonds.**

**11 X 2 = 22**

Tell the your friend to add 2 to the number.

**22 + 2 = 24**

Tell them to multiply that answer by 5.

**24 X 5 = 120**

Ask your friend to add the suit number to the total.

**120 + 2 (diamonds) = 122**

Ask your friend for the answer. When they say "122," secretly subtract 10 from their answer.

**122 - 10 = 112**

Tell them that the card is the Jack of Diamonds. You will know because the Jack equals 11, the first two digits of the total. The last digit is 2, which equals diamonds as on your chart. Wasn't that easy?

# TRICKIER TRICKS

If you enjoyed all the math-a-magical tricks you've learned so far, maybe you're ready to go on to some trickier tricks. These tricks require more practice on the part of the math-a-magician, but they will astound and amaze your friends. Give them a try!

# MENTAL MULTIPLIER

Here's a trick that will make your friends think that you have a calculator built into your head! It is slightly complicated, but the results are amazing. The trick is in finding the 9 compliment, the number you get by subtracting each digit of the number from 9. You will need a calculator for your friend and a piece of paper and a pencil for yourself.

Tell your friend that you can multiply large numbers in your head. Ask your friend to think of any three-digit number.

**Let's say your friend picks the number 674. Write that number down twice.**

**674 674**

Ask your friend to think of another three-digit number.

**Let's say your friend picks 857. Write that number under the number at the left.**

**857**

Finally, add another three-digit number under the number at right. The trick is, when you pick that number, you must subtract each digit of the number on the left from the number 9.

**In this case, your friend's number is 857. Subtract each digit from 9.**

**9 - 8 = 1     9 - 5 = 4     9 - 7 = 2     = 142**

**This is how you find the 9 compliment. The number to use on the right is 142.**

Hand your friend a calculator and tell him or her to multiply the numbers in each column and add the results together.

| | |
|---|---|
| **674** | **674** |
| **X 857** | **X 142** |

While your friend is busy with the calculator multiplying 674 by 857 and 674 by 142, simply subtract 1 from the top number.

**674 - 1 = 673**

Jot that number down and find the 9 compliment by subtracting each digit from 9.

**The number is 673.**

**9 - 6 = 3    9 - 7 = 2    9 - 3 = 6    = 326**

**The 9 compliment is 326.**

Write that number next to 673, and you have the answer minutes before your friend with the calculator.

**The final number is 673,326.**

The secret is finding the 9 compliment.

# COLUMNS OF NUMBERS

Here's another trick using the 9 compliment. In this trick you make your friend do all the work while you appear to solve the problem in seconds. A watch or clock makes this problem more fun.

First you need someone to call out 5 five-digit numbers. This trick works best with a group of people.

Let's say your friends call out these 5 five-digit numbers. Write them in a column.

**18754**

**26979**

**64532**

**29472**

**84754**

Explain that you will add the numbers instantly. To show how hard it is, you will have a friend try to do this trick while you time them. But your friend will need different numbers so as not to give away the answer to the original problem. Here is where you use the 9 compliment. Give your friend a column of numbers where each number is the 9 compliment of the numbers in your column. That is, subtract each digit from 9 to come up with the second column of numbers.

The first number is 18754. Subtract each digit from 9 and write it in a column next to the original numbers. Do this with each number in your column.

**9 - 1 = 8 9 - 8 = 1 9 - 7 = 2 9 - 5 = 4**

**9 - 4 =5**

The first number in your friend's column would be 81245. Your column of numbers is on the left, the numbers that you give your friend are on the right.

| | | |
|---|---|---|
| **18754** | **the nine compliment is:** | **81245** |
| **26979** | **the nine compliment is:** | **73020** |
| **64532** | **the nine compliment is:** | **35467** |
| **29472** | **the nine compliment is:** | **70527** |
| **84754** | **the nine compliment is:** | **15245** |

Have your friend add the second column while you time him or her. Tap your foot and act impatient as they struggle through the addition.

**The answer to the second column of numbers is 275,504. To all this, add 4.**

**275,504 + 4 = 275,508**

Now, amaze your friends by appearing to add the first column in a short time. To do this, find the 9 compliment to the five digits on the right. But first, subtract the digit on the far left from 4. In the example, your friend's total is 275,508. Subtract the leftmost digit (the 2) from 4 and write down the result.

**4-2 = 2**

Then subtract the rest of the digits from 9.

**9 - 7 = 2   9 - 5 = 4   9 - 5 = 4**

**9 - 0 = 9   9 - 8 = 1   = 24,491**

**Put the 2 in front of those digits and you will arrive at the total of the first column.**

**224,491**

The final number is 224,491. This is the total of the numbers that your friends first gave you. You should be able to do this trick in a short time after your friend does all the work to arrive at the answer for the second column.